Everything around us is in motion. We can see planes flying overhead and cars driving down the street. We move as we walk to school or pedal our bikes. Even the planet is moving. But why do things move? And how? These questions have fascinated scientists for centuries.

If you've ever tried to twirl a hula hoop, you know that it's not as easy as it looks! You must move your body in just the right way to set the hula hoop in motion. But why does your moving body move the hula hoop? And why does the hoop eventually stop twirling when you stop moving? The answers to these questions lie in the theories of force and motion.

1

We consider an object to be in *motion* when it is not standing still. This may seem fairly obvious. An object tends to remain at rest or keep moving in a straight line. In order for an object to move or stop moving, a *force* must be applied to it. This is one of the *Laws of Motion*.

A baseball at rest will not move unless someone throws it or a bat hits it. This property of remaining at rest or in motion is called *inertia*. The arm or the baseball bat provides the force that overcomes the ball's inertia.

The tiger's powerful legs exert a force that sets the animal in motion.

Two forces are at work here. The children are applying force to throw the snowballs, and *gravity* is pulling the balls downward. These combined forces are why *projectiles*, such as snowballs, move in an *arc*.

The swan's muscles make its wings flap. The wings move the swan's body, which provides the force that moves the water.

Several forces move this hang glider. The wind keeps the aircraft in motion, while gravity pulls the hang glider downward.

The force that sets an object in motion does one of two things: it either *pushes* or it *pulls*. Pushing and pulling affect how an object moves. At the same time, something is also working against the push or the pull—the object itself! The object exerts an equal force in the opposite direction because of its inertia. This is another law of motion.

As a bulldozer pushes a pile of dirt, the dirt actually applies an equal push in the opposite direction. To get the dirt to move, the bulldozer must apply more force against the dirt than the dirt applies against the bulldozer.

This orangutan moves through the trees by pulling itself. In order for the animal to keep moving, it must continue to pull. Otherwise, it will stop.

ese children are exerting
milar pulling forces against
ch other. This picture illus-
tes what happens when
posite but basically equal
ces cancel each other out.
thing moves!

r a tow truck to
ll an unmoving
r, it must first
ercome the car's
ertia. As the truck
gins to move, the
rce it has to exert
comes less
cause the car
oves and gains
omentum.

The paddle wheel of this boat pushes against the water. As the water is
pushed backward, the boat is pushed forward. This illustrates how a force
(the push) produces an equal force, but in the opposite direction.

Another law of motion explains that forces can *change* an object's motion. For example, a force can speed up motion or slow it down. It can also change the motion's direction, making an object move up and down, back and forth, or around and around.

Many objects move in a straight line. In fact, an object will continue to move in a straight line unless a force acts upon it to change its direction or speed.

Gravity and the cars' engines provide the forces that move the cars down this hill. The curvy road and the cars' brakes add *friction*, which cause the cars to slow. We can move from a straight line into a curvy pattern by changing the force.

As you ride a swing, you experience a very basic motion—back and forth. A back-and-forth movement is called *oscillation,* or *vibration.* The girls' legs provide the force that makes the swing move back and forth.

When you ride on a merry-go-round, you continue to spin, even though you are no longer pushing. This is because the spinning motion has created *angular momentum.*

Objects can also move up and down. In order to move upward, a force must push or pull an object away from the ground. The gears and motor of this escalator pull people upward.

Forces are invisible. But you can see objects that create or transmit forces. You can see the engine of a car. The engine, which is fueled by gasoline, makes different parts of the car move, which in turn rotate the tires, which then push the car along the road. You can't see the wind, although you can feel it. And you can't see the force of a magnet as it pulls and pushes.

Can you see what makes this airplane fly? As the propeller spins, it moves the plane through the air. The air flowing above the wings moves faster than the air moving below the wings. The slow-moving air below the wings keeps the plane in flight.

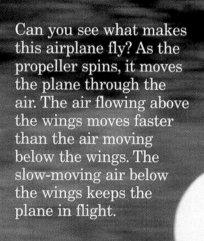

When a bird flies, it uses its wings to keep itself aloft. The back-and-forth movement propels the bird both forward and up and down. By varying the force, the bird can change its speed and direction.

Your feet push the pedals of a bike. This force results in motion you can see. The pedals move gears, which in turn move the wheels, which in turn push you along the street. You can watch all these things move.

Wind exerts a force that sets objects in motion, too. Even though we can't see it, the wind pushes against the sails to make the boat move. The stronger the wind, the faster the boat will travel.

A magnet exerts both a push and a pull. It will pull objects made of iron toward it. It will also pull another magnet. But if the like poles of the magnets face each other, then the forces will push each other away.

Gravity is a force we encounter all the time! It is the force that pulls things to Earth. Gravity is also what keeps the planets spinning through space around the Sun.

Material objects are composed of *mass*. And all mass has *gravity*. Gravity is a force that pulls on, or *attracts,* objects. Because the Sun is massive, it exerts a strong *gravitational pull* on the planets in our solar system.

Just as the Earth moves around the Sun, the Moon moves around the Earth. When astronauts visited the Moon, they had to adjust to its gravity, which is one-sixth that of Earth's. Since the Moon has less mass than Earth, its gravitational pull is less, too.

All the planets in our solar system move, or *revolve*, around the Sun, each in its own path, or *orbit*. The gravitational pull of the Sun keeps all nine planets in their respective orbits. Because each planet has a different mass each has a different amount of gravity.

Nature is full of dramatic examples of movement. Clouds rush across the sky. Tornado winds whirl and spin out of control. Pressure deep inside the Earth pushes up fiery lava. Even waves crashing on a beach have an incredible force.

The Moon's gravitational force pulls on the world's oceans, which create the tides that push and pull at the land. Waves also push and pull, moving back and forth. As they do so, they change the shape of the land, eroding it into amazing formations under the waves' pressure. Wind and tides create waves.

Wind is moving air. A soft breeze is pleasant. But fierce winds, like those from hurricanes, can be destructive. The wind in this picture is pushing and pulling on the palm trees. Extremely strong winds can knock down telephone poles and blow roofs from houses.

Beneath its surface, our Earth is extremely hot. The pressure from this heat builds and builds. When it gets too high, the pressure releases by bursting through a hole in the Earth's crust. This hole is a volcano. The pressure pushes lava, along with ash and steam, from the Earth in a fiery display.

A thunderstorm can be dramatic and beautiful. Lightning is created when a cloud discharges excess electrical energy. Rain falls from the clouds when gravity pulls the drops of water to Earth.

A geyser is similar to a volcano but, instead of lava, water blasts forth. The water comes from hot springs that run below the Earth's surface.

13

People have found ways to control force and motion. No matter how complex, all machines are in some way composed of simple machines. These seven simple machines help us to push and pull and to move objects in various ways.

The wind forces the windmill to turn. The windmill can then apply a force.

Most vehicles include the *wheel and axle*. Wheels helps us overcome friction and move more smoothly and quickly.

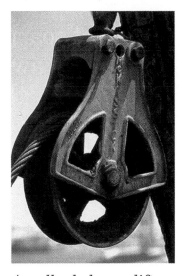

A *pulley* helps us lift things. It magnifies the force we exert.

Gears are wheels with teeth. They are used to increase or decrease force.

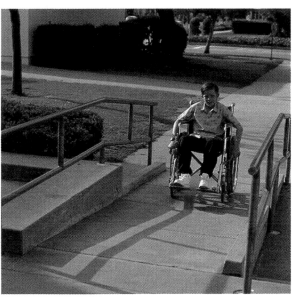

An *inclined plane* lessens effort needed to move an object up or down.

A *screw* is also an inclined plane. It lessens the amount of work because it magnifies the force applied.

A *wedge* is actually two inclined planes back to back. The downward movement creates a sideways force, pushing the wood apart.

A *lever* is a rigid bar fixed at a certain point, like these oars. The rowers move the oars, which are fixed on the boat. The oars push on the water; the water pushes back on the oars— and the boat moves through the water.

Today, our world is full of soaring airplanes, speeding cars, and rockets that push away from Earth. Only by understanding how and why things move, and the forces that set objects in motion, can we create such wonders. By learning how force affects movement, we are able to discover new things about our world—and beyond.

The space shuttle exerts an incredible amount of force to push itself away from Earth. Liquid oxygen mixing with liquid hydrogen provide the energy needed to accomplish this amazing feat.